WE CAN READ about NATURE!™

WETLANDS

by CATHERINE NICHOLS

BENCHMARK BOOKS

MARSHALL CAVENDISH
NEW YORK

With thanks to
Peggy C. Hansen, Teacher,
Noxon Road Elementary School,
New York, for providing the activities in
the Fun with Phonics section and
to Beth Walker Gambro, Reading Consultant.

Benchmark Books
Marshall Cavendish
99 White Plains Road
Tarrytown, New York 10591-9001

Photo Research by Candlepants Incorporated

Cover Photo: Corbis / Tom Bean

The photographs in this book are used by permission and through the courtesy of;
Corbis : Tom Bean, 4-5, 6; Tim Wright, 7; Dave G. Houser, 8-9;
Phil Schermeister, 10, 16-17; Peter Finger, 11; Tony Arruza, 12-13; David Muench, 14;
Wayne Lawler, 15, 26; Bill Ross, 18; Hal Horwitz, 19; Joe McDonals, 20, 21(bottom);
Bryn Colton, Assignments Photographers, 21(top); Lynda Richardson, 22, 23;
Buddy Mays, 24; Scott T. Smith, 25; Robert Holmes, 27; Layne Kennedy, 28-29.

Library of Congress Cataloging-in-Publication Data

Nichols, Catherine.
Wetlands / by Catherine Nichols.
p. cm. — (We can read about nature!)
Includes index.
Summary: Presents the different kinds of wetlands, such as swamps, marshes, and bogs,
and introduces some of the animals that live there.
ISBN 0-7614-1434-7
1. Wetlands—Juvenile literature. [1. Wetlands. 2. Wetland ecology.
3. Ecology.] I. Title. II. Series.
QH87.3 .N53 2002
577.68—dc21
2002003781

Printed in Hong Kong

1 3 5 6 4 2

Look for us inside this book.

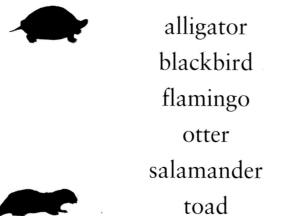

alligator
blackbird
flamingo
otter
salamander
toad
turtle

Land that is covered by water is called a wetland.

Plants grow in the water and in the wet soil.

A wetland can be as small as your backyard.

Some wetlands are as big as cities.

Everglades National Park

A marsh is one kind of wetland.

Trees cannot grow in a marsh.

It is too wet there.

Grassy plants can grow in a marsh.

Cattails

Water lilies can too.

A swamp is another kind of wetland.

It is like a wet forest.

Swamps have trees.

Cypress trees

Their roots stick out above the water.

Mangrove trees

Bogs are wetlands that are found in cool places.

A bog soaks up rainwater.

The ground feels soft and springy.

Mosses grow in bogs.

These special plants also grow there.

Venus's-flytrap

What makes them special?

They eat insects!

Pitcher plants

All kinds of animals live in wetlands.
There are turtles, otters, and even
alligators.

Turtle

Otters

Alligator

Some animals can live on land and in water.

Toad

Spotted salamander

There are many birds in wetlands.

Flamingo

Yellow-headed blackbird 25

Some birds build nests there.

Other birds just stop by to rest.

Wetlands are not just for animals.

They are for all of us to enjoy.

Fun with Phonics

(Answers on page 32)

1. TREASURE HUNT!

A. Find a word on page 5 that rhymes with slow.
B. Find two compound words (words made up of two complete words, such as butterfly) on page 6.
C. Find a word on page 9 that rhymes with pet.
D. Find a word on page 10 that starts out like play.
E. Find a three-syllable word on page 12.
F. What word ends like block on page 15?
G. Find three words on page 17 with the hard g sound (the g sound in goat).
H. What word rhymes with finds on page 20?
 I. Find three words on page 20 with double consonants.
J. Find three words on page 27 with st.
K. Find a word on page 29 with the vowel sound in boy.

2. WHICH WETLAND?

A.
1. This is one kind of wetland.
2. It has five letters.
3. It is too wet for trees to grow.
4. It is the first syllable of a sweet food that is sometimes toasted.
5. It ends with two letters that mean "Be quiet."

B.
1. This is one kind of wetland.
2. It has five letters.

3. Trees grow here.
4. It starts out like swing.
5. It ends like camp.

3. CHALLENGE: PLURALS

Plural means more than one. Write the plural form of these words on a separate paper. (The plural words are all in the story, so look back if you need help!)

plant	swamp	insect
city	root	turtle
tree	moss	otter
lily	bog	nest

Fun Facts

- The Everglades in Florida is a big freshwater marsh. It covers about 7,500 square miles.
- The roots on cypress trees are called "knees." These knees stick up above the water. They hold the tree steady.
- The Venus's-flytrap and pitcher plant grow in wetlands. The Venus's-flytrap has sticky leaves that catch insects. Pitcher plants drown insects in their long water-filled tubes.
- Alligators can grow very big in a wetland. One captured alligator was almost 14 feet long. It weighed more than one thousand pounds!

Glossary/Index

About the Author

Catherine Nichols has written nonfiction for young readers for fifteen years. She works as an editor for a small publishing company. She has also taught high-school English. Ms. Nichols lives in Jersey City, New Jersey, with her husband, daughter, cat, and dog.

Answers to pages 30–31:

1. Treasure Hunt!
A. grow B. wetland, backyard C. wet D. plants E. another F. stick G. bog, ground, springy H. kinds I. all, otters, alligators J. just, stop, rest K. enjoy
2. Which Wetland? A. marsh B. swamp
3. Challenge: Plurals plants, cities, trees, lilies, swamps, roots, mosses, bogs, insects, turtles, otters, nests